TEAM EARTH

SYMBIOTIC
RELATIONSHIPS
ANIMALS AND PLANTS
WORKING TOGETHER

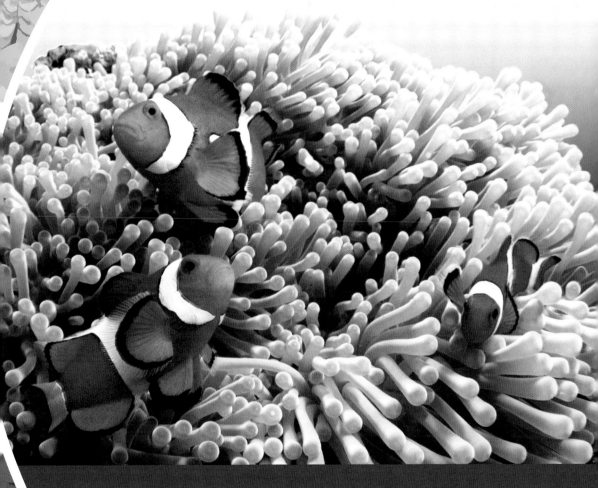

BY EMMA HUDDLESTON

CONTENT CONSULTANT
Michele Nishiguchi, PhD
Department Head, Department of Biology
New Mexico State University

Cover image: Clown fish and sea anemones live
together in a symbiotic relationship.

Core Library

An Imprint of Abdo Publishing
abdobooks.com

abdobooks.com

Published by Abdo Publishing, a division of ABDO, PO Box 398166, Minneapolis, Minnesota 55439. Copyright © 2020 by Abdo Consulting Group, Inc. International copyrights reserved in all countries. No part of this book may be reproduced in any form without written permission from the publisher. Core Library™ is a trademark and logo of Abdo Publishing.

Printed in the United States of America, North Mankato, Minnesota
092019
012020

THIS BOOK CONTAINS RECYCLED MATERIALS

Cover Photo: Shutterstock Images
Interior Photos: Shutterstock Images, 1, 9, 10, 30 (grass), 30 (mouse), 30 (raccoon), 30 (bobcat); Maurizio Lanini/iStockphoto, 4–5; Paula French/iStockphoto, 7; Magnus Larsson/iStockphoto, 12–13; Henk Bogaard/Shutterstock Images, 14, 43; William Sherman/iStockphoto, 18–19; Francois Gohier/Science Source, 22–23; Fiona Ayerst/Shutterstock Images, 24, 45; Sinclair Stammers/Science Source, 26–27; Nigel Cattlin/Science Source, 29; Andrew Krasovitckii/Shutterstock Images, 30 (snake); Kateryna Kon/Shutterstock Images, 34–35; Danté Fenolio/Science Source, 38

Editor: Marie Pearson
Series Designer: Megan Ellis

Library of Congress Control Number: 2019942107

Publisher's Cataloging-in-Publication Data

Names: Huddleston, Emma, author.
Title: Symbiotic relationships: animals and plants working together / by Emma Huddleston
Other Title: animals and plants working together
Description: Minneapolis, Minnesota : Abdo Publishing, 2020 | Series: Team earth | Includes online resources and index.
Identifiers: ISBN 9781532191022 (lib. bdg.) | ISBN 9781644943298 (pbk.) | ISBN 9781532176876 (ebook)
Subjects: LCSH: Symbiosis--Juvenile literature. | Animal-plant relationships--Juvenile literature. | Pollination--Juvenile literature. | Herbivores--Juvenile literature. | Plant-microbe relationships--Juvenile literature.
Classification: DDC 577.85--dc23

CONTENTS

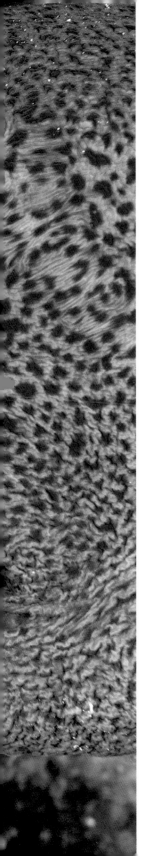

LIVING TOGETHER

A moray eel swims in shallow water. It lives in tropical waters of the Indian and Pacific Oceans. It is 9 feet (2.7 m) long. Its thick skin is greenish brown. Its sharp teeth look dangerous. But a bluestreak cleaner wrasse is not afraid. Wrasses are tiny cleaner fish. It waits on rocks nearby. It is not scared of any larger sea creatures.

The wrasse waits for the eel to stop swimming. The eel slows down. The little fish jumps into action. It inches over the eel's body. It goes into the eel's mouth. It doesn't have to worry about being eaten. Instead, this is its time to eat. It eats parasites and dead skin

Cleaner fish such as bluestreak cleaner wrasses help keep other sea creatures healthy.

SPECIES CLASSIFICATION

Scientists have a method for organizing living things called classification. Classification helps show how species are or are not related to one another. The chart for classification is an upside-down triangle. Domain is the broadest level. Each domain is divided into kingdoms. From there, the classification gets more and more specific. The levels get smaller from kingdom to phylum, class, order, family, genus, and species. Species is the most specific. A species is a group of animals that have similar traits and can interbreed. Gray wolves, koalas, and African bush elephants are all examples of different species.

off the eel. The eel benefits because it gets cleaned. One wrasse can do 2,000 cleanings in one day.

The eel and cleaner fish might seem unlikely partners. They are an example of a symbiotic relationship. Symbiosis is when two species live together. The species are not alike, but they have a close relationship.

Some species only interact with each other occasionally, such as at a watering hole. They do not necessarily have a symbiotic relationship.

CLOSE INTERACTIONS

Animals and plants interact constantly. They come in contact with one another in their habitats.

Each environment has many species. A jungle has green leafy plants. Monkeys and colorful birds might call it home. A desert has cacti. Beetles and snakes might live there. Many gardens have flowers and insects. Different species live near each other or on one another. Some share food sources or living space.

In 2017, scientists found a fossil in China.

HOW PLANTS AND FUNGI GROW TOGETHER

One type of symbiotic relationship happens in 80 percent of plants. A fungus attaches to a plant's roots. The fungus brings additional nutrients to the plant from the soil. This helps the plant grow. The plant is food for the fungus. Both species benefit. Plants and fungi formed partnerships nearly 1 billion years ago. Plants used to only grow in water. Fungi helped plants start to grow on land. Their symbiotic relationship helped them grow nearly everywhere on land. It also helped them survive over time.

PARTS OF AN ECOSYSTEM

An ecosystem is made of living and nonliving parts. Nonliving parts include sunlight, air, water, and rocks. They support species that live there. The living parts include bacteria, plants, and animals. Can you find two examples of both parts of an ecosystem in this image?

It showed evidence of a symbiotic relationship. The fossil was a sea worm from 520 million years ago. Smaller worm-like species were attached to it. It shows that species have been living together for a long time.

The anemone hermit crab has a symbiotic relationship with sea anemones. The anemones grow on its shell.

SYMBIOTIC RELATIONSHIPS

The three main types of symbiotic relationships are mutualism, commensalism, and parasitism. Each type of relationship is beneficial to at least one party. Mutual relationships are good for both species. Commensal relationships benefit one or neither of the species, but they don't do any harm either. Parasitic relationships help one species and harm the other.

Symbiotic relationships are a key part of the natural world. They make a difference in small ways. One example is by exposing a species to others that they have nothing in common with. When species come together, they can benefit in unexpected ways.

They work together or get a food source that they might otherwise have no access to. Symbiotic relationships also provide humans with food and help trees and plants grow.

Symbiotic relationships create balance. Bumblebees help flowering plants by spreading pollen. In exchange, the plant's flowers are a food source for the bees. Without their partner, each species would struggle. The number of living individuals could go down. Plants, animals, and humans depend on symbiotic relationships in daily life.

EXPLORE ONLINE

Chapter One gives an introduction to symbiotic relationships. They impact life on Earth in big and small ways. As you know, each source is different. Explore this website about the three main types of symbiotic relationships. What new information does it explain?

HOW WE WORK TOGETHER: SYMBIOSIS EXPLAINED

abdocorelibrary.com/symbiotic-relationships

CHAPTER
TWO

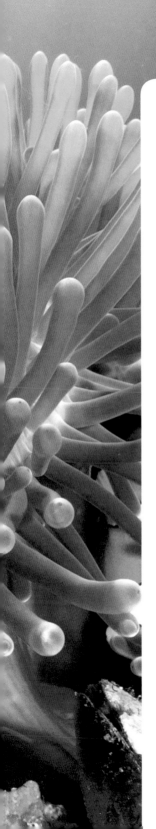

MUTUALISM

One example of mutualism, where both parties benefit, is the relationship between sea anemones and clown fish. Sea anemones are dangerous to most small ocean animals. Their long tentacles sting anything that swims into them. But sea anemones do not sting clown fish. Clown fish have a thick layer of mucus on their skin. The mucus is a sticky protective covering. It stops the sea anemone from recognizing them as food. Instead, clown fish use sea anemones for shelter. They live in the sea anemones. The anemones protect the fish.

This partnership benefits sea anemones because the clown fish scare off larger fish that

Clown fish and sea anemones both benefit from their relationship.

Oxpeckers keep animals such as rhinoceroses free of pesky bugs.

might eat them. The movement of the clown fish also helps the sea anemone breathe. It stirs up the water flow, bringing fresh water to the area.

Mutual relationships help both species, but they are complicated. In order for both species to benefit, their relationship has to be balanced just right.

ANIMALS HELPING EACH OTHER

Oxpeckers are birds with bright yellow-and-red beaks. They ride on the backs of large animals in Africa.

Some animals they ride on are the giraffe, rhinoceros, and elephant. Most of the time, this relationship is mutual. The oxpecker gets free rides and food. They eat ticks and insects off an animal's skin. One oxpecker can eat up to 400 ticks a day. This benefits the animal because ticks are annoying, and they can carry disease. The birds also warn the animals when danger is near. They fly up and squawk.

However, the relationship can become unbalanced. Oxpeckers can benefit at the expense of the larger animal. Groups of oxpeckers can be a burden to carry from place to place. Sometimes they pick at wounds on an animal's skin.

HOW DO PLANTS BENEFIT?

Plants need symbiotic relationships to thrive.

DIGESTING WOOD

Wood is the main food source for beetles and termites, but it is not easy to break down. Both species need help from microbes in a symbiotic relationship. Microbes are tiny bacteria. They live in the insects' guts and help them digest.

Honeybees are pollinators. Their furry legs and bodies collect pollen as they drink nectar from a flower. They carry the pollen to another flower. This pollination enables plants to make seeds. The bees and flowers have a mutual relationship. Each species benefits.

OBLIGATE MUTUALISM

In some cases, two species have to work together. Without their symbiotic relationship, they can't survive. This is called obligate mutualism. In the ocean, coral reefs have a mutual relationship with algae called zooxanthellae. They rely on zooxanthellae to survive. The algae produce oxygen. Oxygen is a gas that helps humans and animals breathe. The algae also break down waste into nutrients for the corals. In exchange, the corals provide a place for the algae to live. The corals use substances from the algae to make food. Their relationship is a cycle. Both help create nutrients for the other.

Sometimes corals get stressed. This often happens when water temperatures are too warm. When corals are stressed, they end their relationship with zooxanthellae by releasing them. This causes corals to bleach and turn white. The algae are what gave the corals a vibrant color. Bleaching can lead to death because corals need algae to survive.

FURTHER EVIDENCE

Chapter Two discusses mutual relationships. Review the chapter. Identify the main point. The website below also describes mutual relationships. How is the information similar? How is it different?

EXAMPLES OF MUTUAL RELATIONSHIPS

abdocorelibrary.com/symbiotic-relationships

CHAPTER
THREE

COMMENSALISM

Another type of symbiotic relationship is commensalism. This is a partnership where one species benefits. The other species is barely affected, if at all. Many commensal relationships happen for the purpose of food or movement. *Commensalism* means "eating at the same table." This is why relationships where two species eat in the same place or at the same time are called this.

EATING AT THE SAME TABLE

Wingless mites attach to bees and butterflies. They get a ride to different flowers. This helps the tiny mites get food. The other insects

Cattle egrets may perch on horses and ponies as the birds watch for insects.

WHY DO SLOTHS POOP ON THE GROUND?

Sloths live in Central and South America. They risk their lives climbing down from their trees to poop. Larger animals can easily attack them. Scientists discovered sloths may do this because of a symbiotic relationship. One species of moth lives in a sloth's fur. While the sloth poops, the moths lay their eggs in the fresh stool. That is where young moths grow into adults. In return, moths bring nutrients to the sloth's fur to help algae grow. Algae is a food source for sloths. It grows in sloths' fur. But it grows best on sloths that also have moths in their fur.

are not harmed. Several tiny insect species benefit from relationships with ants and termites. They live in anthills and termite mounds. This gives them shelter and access to leftover food. They have little or no effect on the ants and termites.

Cattle egrets are birds that eat insects in grasslands. They live in open fields and wetlands around the world. They hunt near cattle and horses because the large animals stir up the grass. This causes insects to fly and move. The egrets can easily find food. Their presence doesn't change the way the cattle and horses live.

Barnacles are small ocean animals with hard shells. Acorn barnacles are the most common species. They live in cold northern waters of the Atlantic and Pacific Oceans. Barnacles' shells develop over their lifetime. First they start out as tiny larvae. They float through the water until they find a surface to settle on. The underside of a whale is a perfect location. Usually barnacles attach to the whale's head or fins.

ENDANGERED RELATIONSHIPS

A species is endangered when it is at risk of dying off forever. Many whale species are endangered. Water around the world is rising in temperature. This threatens whales' food sources. Warming waters also affect where whales swim. When one species is endangered, all the other animals and plants they have a relationship with are at risk too. Without whales, barnacles would not have bodies to attach to and spread to new habitats.

Barnacles often grow on gray whales.

These places have constant water flow. They also put the barnacle close to food. Both species eat at the same time. Whales swim through water with small fish and tiny shrimp-like creatures called krill. Barnacles take in the tiny creatures and nutrients from the water.

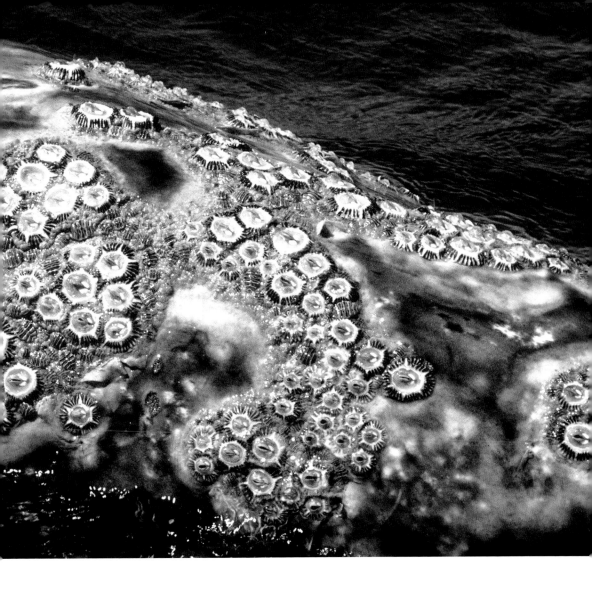

A humpback whale can be up to 62 feet (19 m) long and weigh up to 80,000 pounds (36,200 kg). One humpback whale can support 1,000 pounds (454 kg) of barnacles. It does not benefit from its relationship with barnacles, but the barnacles do not hurt it.

Suckerfish do not hurt the animals they attach to, though they can cause irritation.

WHAT TYPE OF RELATIONSHIP IS IT?

Commensal relationships are less common than other types of symbiotic relationships. This is because it is difficult for one species to be in a relationship with another and not affect it. Usually they experience a small amount of benefit or harm.

Remoras are fish that live in warm, tropical waters. They are also known as suckerfish. This is because a suckerfish has a large, flat suction cup on its head. It helps the suckerfish attach to the bodies of larger fish. Suckerfish often attach to whale sharks and other large ocean animals.

They do this to get a free ride to food. After the shark attacks and eats prey, the suckerfish detach. They eat the leftover scraps. Then they attach again. This seems like commensalism. However, sometimes suckerfish clean the shark's mouth too. This could make the relationship mutual.

In small numbers, the suckerfish don't cause many problems. But in larger numbers, they can affect how quickly the shark swims. Sometimes they also irritate the shark's skin. For these reasons, some people wonder if the relationship harms the shark.

PARASITISM AND PREDATION

Some commensal relationships turn into parasitic relationships. This happens often with wasps. Wasps are parasites. They live around the world. Parasites are species that survive by living on or inside another living thing called the host. For example, one wasp species attaches to female white butterflies. This relationship is not a problem for the butterfly because it just carries the wasp to a new location. However, once the butterfly lays eggs, things change. The wasp has a parasitic relationship with the butterfly eggs. It lays its own larvae

The *Trichogramma brassicae* wasp lays its larvae on the eggs of a cabbage white butterfly.

BRAINWASHED CATERPILLAR

One species of wasp is a unique parasite. It injects up to 80 larvae into a caterpillar. The larvae grow by eating the host's blood and body fluids. They make sure not to kill it. Then, the larvae chew their way out of the caterpillar's body. At the same time, they release a chemical. Scientists believe this chemical changes the caterpillar's behavior. Once the larvae are free, the brainwashed caterpillar helps spin a cocoon around the growing wasps instead of itself. Then, it guards and protects the wasps until it dies.

inside them. The wasp larvae eat the butterfly eggs to grow and survive.

PARASITES HELPING OUT

Aphids are known as plant pests. They are tiny insects with soft bodies. They can do serious harm to plants. They carry disease and cause deformed leaves, buds, and flowers. In the United States, sometimes thousands of aphids attack a tomato field. They eat and kill the tomato plants that humans are trying to grow for food.

The black wasp larvae leave the shells of the aphids behind. Their work is visible by a hole left in the back end of the aphids.

The black wasp is a parasite that can help. It acts as natural pest control. The wasp carefully injects one larva into each aphid. One black wasp can do this to 200 aphids. Then, the wasp larvae eventually kill the aphid by eating it from the inside out. The black wasp is a parasite, but it is also known as a beneficial insect.

CHAINS

A food chain is the order in which living things get eaten in an ecosystem. This woodland food chain shows the order in which some animals get eaten in a forest. Think about another environment such as the desert or ocean. What plants, insects, and animals would be in that food chain?

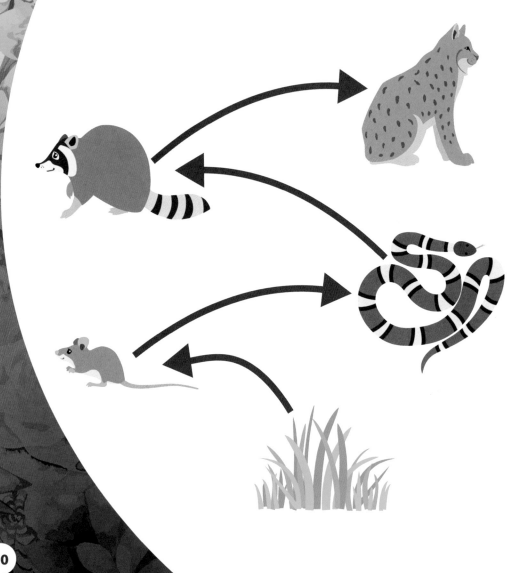

Beneficial insects help humans and the environment. They protect plants, and plants are a basic food source for many species, including humans. Without beneficial insects, life on Earth would struggle. Pest species would not be controlled by natural predators. Plants would not have help with pollination. Additionally, dead animal and plant matter would pile up. Insects play a big role in helping break down waste.

PREDATION

Predation happens when one species eats

TAPEWORMS

Tapeworms are parasites. They are flat, long creatures. They live in humans and animals. They have a hook or sucker on their front end. It helps them stay in place inside their host's intestines. Tapeworms feed off food from their host. They take in nutrients as the host digests food. Tapeworms in humans can grow up to 60 feet (20 m) long. They are known for causing problems. Their hosts eventually die from lack of nutrients. However, a recent study found one species of tapeworm that may be helpful. Rats with this tapeworm were better protected from memory loss compared to rats without it.

another to survive. This type of relationship is common between insects and plants. Plants are rooted in one place and cannot get away. Insects take advantage of that. Many eat plants. Locusts are an extreme example. They do not have one specific plant species that they eat. They will eat many types of green plants. This harms and sometimes kills plants. Locusts live around the world. But they are known for their negative impact on crops in Africa. The desert locust lives in Africa, the Middle East, and Asia. A swarm of desert locusts can eat up to 423 million pounds (192 million kg) of plants in a day.

Predator and prey relationships are a type of symbiosis. The predator benefits. It needs the prey to survive. It is different from a parasitic relationship because predators do not live in or on the prey. Instead, predators usually sit above prey in the food chain.

STRAIGHT TO THE
SOURCE

Angela Douglas is a biologist who studies insects. Douglas believes learning about how insects and plants relate can lead to better pest control solutions.

There's something special about insects that enable them to [eat plants in all stages of the insects' lives]. They cause major damage in crop production.

We are currently working on developing alternative approaches to disrupt the interaction between the animal, the insect, and the symbiotic bacteria as a novel pest control strategy against these insects. This is really important, I think. It's important because traditional chemical insecticides are posing increasing problems. There's increasing incidents of resistance [and] an increasing regulatory burden . . . because of environmental and human health issues.

Source: "Angela Douglas, Cornell University: Host-Microbe Interactions Affect Metabolism & Nutrition." *Christian-Albrechts-Universität zu Kiel*. YouTube, June 13, 2017. Web. Accessed July 22, 2019.

What's the Big Idea?
Read the quote carefully. What is its main idea? Name two or three details that support the main idea, and explain how they support it.

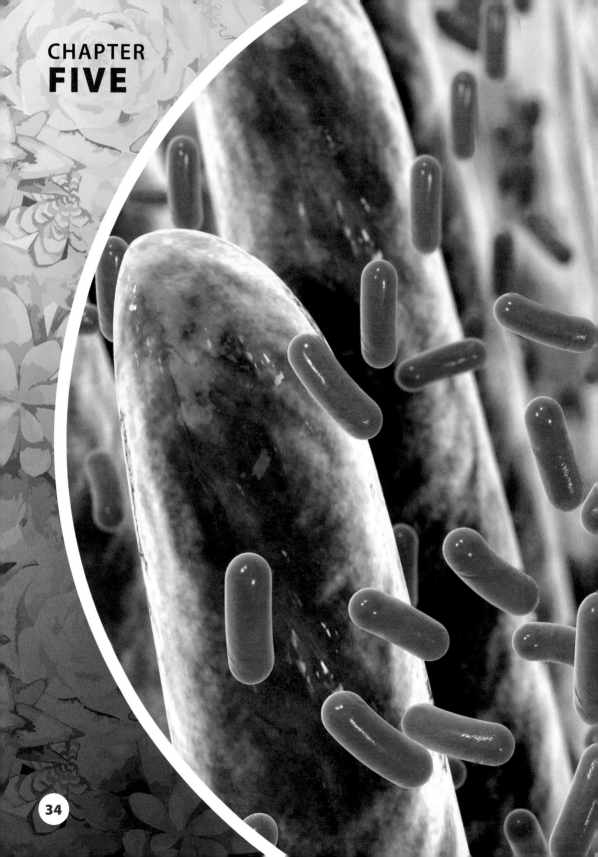

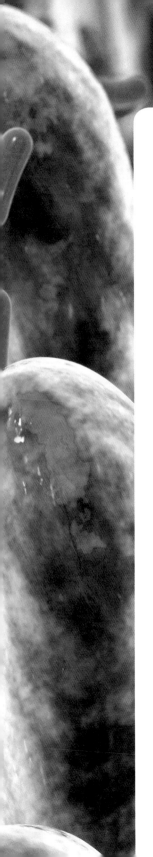

EVOLUTION

Some bacteria live inside organs of the human body. They help digest food. Scientists have studied this symbiosis. As the human body changes or eats different things, the bacteria have to adapt, or change in order to perform better. They evolve to survive. At the same time, they continue to help the human during digestion.

Evolution is a process of gradual change. It happens so a species can survive. Predation can lead to evolution. It causes prey to adapt. Prey need to avoid being eaten, so they adapt gradually. Traits that help a species survive are passed on to the next generation. Over

Bacteria, *red*, in the intestines are important for helping digest food.

PASSING ON TRAITS

Evolution affects traits passed from one generation to the next. Mimosa plants have a unique behavior. They close their leaves upon touch in case of potential danger. Other plants do not have this ability. Deer with large antlers attract mates better than deer with small antlers. This trait is passed on when the deer mate. The walking stick and walking leaf are two insect species. Their color and size look like sticks and leaves. This helps them hide from predators.

time, unhelpful traits disappear in the process of evolution.

WORKING TOGETHER TO SURVIVE

Ocean life is affected by global warming. Global warming is the long-term rising temperature of Earth. Scientists think symbiosis might be the solution. Species can work together to survive in changing environments. For example, a species of sponge is thriving in warmer waters because of the algae that lives on it. The algae help produce nutrients for the sponge.

In 2018, scientists studied anglerfish from the Gulf of Mexico. Anglerfish live in deep, dark waters.

Females have a long organ that comes out of their forehead. Glowing bacteria live in the bulb on the end. This creates a mutual relationship. The anglerfish provide nutrients for the bacteria, and the bacteria give off light. Light attracts prey and potential mates to the fish.

This species of bacteria can live in two locations. It can live inside the anglerfish or float freely in the water. This study showed that this species of bacteria is not done evolving. It is still changing based on its environment. The bacteria living in

WHAT ARE LICHENS?

Lichens are an example of a symbiotic relationship. Their mass is 90–95 percent fungus. For this reason, they are classified based on the fungus species in the relationship. The rest of the mass is an alga, bacteria species, or both. All of the partners together create a lichen. More than 15,000 species of lichens exist. Lichens can thrive in harsh environments. They can live in extreme heat and cold. They can grow on bare rock, lava flows, and dead wood. Most living things can't survive in these places. But a lichen's symbiotic relationship is made to survive.

Anglerfish are named because of how females use a glowing bulb that grows out from their heads to attract prey.

anglerfish adapted. They lost genes related to digestion because they can rely on the fish to supply them with nutrients. Free-living bacteria from this species kept these genes. They are still able to live on their own.

A CASE OF COEVOLUTION IN THE DESERT?

Most pollinators accidentally spread pollen. It brushes on and off their body as they eat nectar from different flowers and plants. Joshua trees have waxy flowers and grow in the Mojave Desert in California. Most insects do not land on them to eat. The trees also have spiny leaves and branches that can be unsafe to touch. Joshua trees have to rely on one insect species to help them pollinate.

Yucca moths are tiny gray moths that pollinate Joshua trees. The trees' seedpods are the only food source for yucca moth caterpillars. This is why they help the tree. The moths use their jaws to collect pollen and spread it to other plants. Along the way, they lay eggs in the seedpods. Each species helps the other reproduce.

Scientists think the moths and Joshua trees are an example of coevolution. Coevolution is when two separate species evolve at the same time. They make changes based on each other. There are two species of Joshua tree. One is shorter than the other. The yucca moth also has two species. One has a shorter body than the other.

In 2014, scientists discovered that large moths that landed on Joshua trees with small flowers could damage the flowers. Small moths that pollinated larger Joshua trees almost never laid their eggs successfully in the seedpods. Each of the species grew and survived best when working with the species of matching size.

SYMBIOSIS BEYOND NATURE

Humans sit at the very top of the food chain. We eat many types of plants, insects, and animals. We have different ways of growing, catching, and hunting food. Human activity threatens many species. Grizzly bears have been overhunted. Whales are accidentally injured or killed by boats. Trees are cut down to make wood and paper. Plant life is destroyed to make space for new buildings and roads. Today, scientists and environmentalists look for ways humans can live in harmony with nature.

STRAIGHT TO THE
SOURCE

Peter Laybourn began an industrial symbiosis program in Europe. Industrial symbiosis is when companies share materials, energy, or information. Laybourn explains the benefits:

> *If we organize ourselves in the industrial world more like nature then we should be more sustainable.*
>
> *We take this metaphor and use it in industrial symbiosis by encouraging traditionally separate industries to work together for mutual benefit. Basically it is rather like birds sitting on the back of a cow and eating its ticks. . . .*
>
> *We have already saved industry more than €140 million [$215 million] and in doing so diverted more 3.9 million tonnes [4.3 million tons] of waste away from landfill, and reduced CO2 emissions by 4.9 million tonnes [5.4 million tons].*
>
> Source: "Working Together to Boost Industrial Sustainability." *Eco-Innovation Action Plan*. European Commission, October 30, 2008. Web. Accessed July 22, 2019.

Back It Up

Write a paragraph describing the benefits of industrial symbiosis in your own words. Then write down two or three pieces of evidence from this passage.

FAST FACTS

- A symbiotic relationship is when two species live together. They interact closely.

- The three main types of symbiotic relationships are mutualism, commensalism, and parasitism. Each type of relationship benefits one or both species.

- Mutual relationships benefit both species. Sometimes two species, such as corals and zooxanthellae, have to work together, or the corals won't survive.

- Commensal relationships benefit one species and barely affect the other species. They often occur for the purpose of food or movement.

- Parasitism is a relationship where one species benefits at the expense of the other. Parasites such as wasps are species that survive by living on or in a host.

- Parasites can act as natural pest controls. This is an example of a beneficial insect. Beneficial insects help humans and the environment.

- The relationship between predators and prey is part of the food chain cycle. Energy and matter move from one living thing to the next, usually by being eaten.

- Evolution is the process of gradual change. Species evolve in order to survive or live a better life. Helpful traits get passed from one generation to the next. All species are constantly evolving in response to their environment.

- Species can work together to survive in changing environments. Algae help sponges live in warmer waters.

- Humans can use symbiosis as a model for working together.

STOP AND
THINK

Take a Stand

Chapter Five discusses how human actions negatively affect nature. Some people think humans need to change the way they live in order to have a mutually symbiotic relationship with nature. Other people don't think that humans need to change their actions very much because people are different from plants, insects, and animals. What do you think humans' relationship with nature should be like? Explain whether or not it fits perfectly into one of the three categories: mutualism, commensalism, parasitism. Use examples to support your opinion.

Why Do I Care?

Maybe you have never seen a symbiotic relationship in nature. That doesn't mean you can't think about how they affect the world around you. How does symbiosis in nature affect your life? What role do bacteria, plants, insects, and animals play in your life? How might your life be different without them?

Another View

This book talks about scientists studying different species. As you know, every source is different. Ask a librarian or another adult to help you find another source that talks about scientific studies. Write a short essay comparing and contrasting the new source with this book. What is the main point of the study? What is the main point of this book? How are they similar and why? How are they different and why?

You Are There

This book discusses studies about Joshua trees in the desert and coral reefs in the ocean. Imagine you are traveling with a scientist to study in one of these locations. Write a letter home telling your friends what you have found. What do you notice about the different species and their relationship? Be sure to add plenty of detail to your notes.

GLOSSARY

alga
a plant-like living being

environmentalist
a person who is concerned
about Earth and takes action
to help it

gene
a tiny unit that carries
the traits that make up
living beings

insecticide
a chemical substance used to
kill insects

larva
the early form of an animal
that usually hatches from
an egg

parasite
a species that survives by
living on or in a host species

pollination
the process of transferring
pollen from one plant
to another

regulatory
related to laws or rules

resistance
a species' ability to survive
despite coming in contact
with harmful substances

species
a specific kind of living being

trait
a physical feature or behavior
received from parents

ONLINE RESOURCES

To learn more about symbiotic relationships, visit our free resource websites below.

Visit **abdocorelibrary.com** or scan this QR code for free Common Core resources for teachers and students, including vetted activities, multimedia, and booklinks, for deeper subject comprehension.

Visit **abdobooklinks.com** or scan this QR code for free additional online weblinks for further learning. These links are routinely monitored and updated to provide the most current information available.

LEARN MORE

Hamilton, S. L. *Bees & Wasps*. Minneapolis, MN: Abdo Publishing, 2015.

Kalman, Bobbie. *Symbiosis*. New York: Crabtree Publishing, 2016.

INDEX

About the Author

Emma Huddleston lives in the Twin Cities with her husband.
She enjoys writing educational books, but she likes reading
novels even more. When she is not writing or reading, she
likes to stay active by running and swing dancing.